Ilson Rodrigues dos Reis Junior
Flávio S. Amâncio Jr.
Luis Fernando A. Araújo

Evaluation of the gravel reserves of regions in the municipality of Araxá-mg

Ilson Rodrigues dos Reis Junior
Flávio S. Amâncio Jr.
Luis Fernando A. Araújo

Evaluation of the gravel reserves of regions in the municipality of Araxá-mg

The growing use of gravel

Imprint

Any brand names and product names mentioned in this book are subject to trademark, brand or patent protection and are trademarks or registered trademarks of their respective holders. The use of brand names, product names, common names, trade names, product descriptions etc. even without a particular marking in this work is in no way to be construed to mean that such names may be regarded as unrestricted in respect of trademark and brand protection legislation and could thus be used by anyone.

Cover image: www.ingimage.com

This book is a translation from the original published under ISBN 978-613-9-72406-2.

Publisher:
Sciencia Scripts
is a trademark of
Dodo Books Indian Ocean Ltd. and OmniScriptum S.R.L publishing group

120 High Road, East Finchley, London, N2 9ED, United Kingdom
Str. Armeneasca 28/1, office 1, Chisinau MD-2012, Republic of Moldova, Europe
Printed at: see last page
ISBN: 978-620-7-92719-7

AUTHORS' BIOGRAPHIES

Ilson Rodrigues dos Reis Júnior

Born in Araxá (MG) in 1995. Graduating in Civil Engineering at the Araxá Planalto University Centre - 2018.

Flávio Sérgio Amâncio Júnior

Born in Araxá (MG) in 1996. Graduating in Civil Engineering at the Araxá Planalto University Centre - 2018.

Luis Fernando Aguiar Araújo

Born in Araxá (MG) in 1996. Graduating in Civil Engineering at the Araxá Planalto University Centre - 2018.

Diogo Aristóteles Rodrigues Gonçalves

Born in Araxá (MG) in 1986. Graduated in Agronomy at FAZU -2009-, post-graduated in University Teaching at the Centro Universitário do Planalto de Araxá -2015-, Master in Plant Production at UFV/CRP - 2014- and PhD student in Agronomy at UFU/MG.

SUMMARY

The municipality of Araxá-MG has an extensive road network in a poor state of repair, so there is a need to carry out studies to help locate materials that can help improve this sector.

Nowadays, with the growing use of gravel in paving, motivated by the high rate of urbanisation and repairs to rural roads, there is a tendency to opt for extracting this material in such a way as to obtain a good end result at a low cost.

Due to the increased weight of the means of transport, the use of more resistant methods for paving has also increased, so the demand for gravel for this use has become greater.

With the incessant increase in demand, gravel is becoming harder to findevery day.

Consequently, we are looking for new sources of this material, which are easy to access, low cost and with basic standards that meet your application requirements.

The aim of this work is to evaluate the gravel reserve(s) in the Serrinha, Itaipu and Marmelo regions, located in the municipality of Araxá-MG.

The study was carried out through field sampling and laboratory tests. Results were obtained that qualify and quantify the gravel deposit.

Keywords: Gravel. Paving. Maintenance. Quarry.

SUMMARY

CHAPTER 1

INTRODUCTION

With the economy booming and the country's main means of transporting its production being the roads, large stretches of streets, avenues and motorways are created, renovated and expanded every day.

With this high demand for aggregates for these stretches, there is a need to use different materials due to the scarcity of materials commonly used to date (OLIVEIRA, 2012).

Gravel, also called boulders, is an unconsolidated river sediment made from igneous rock, made up of grains with a general diameter of more than 5 mm, with larger grains reaching diameters of up to 100 mm. According to MARANGON (2009), gravel is used for surfacing roadbeds, building earth embankments, concrete, drainage works, among others.

Also genetically known as pebbles, gravel originates from pre-existing rock fragments and falls within a determining and characteristic granulometric range. It is an aggregate of natural origin and large size.

Gravel belongs to the Construction Aggregates group (sand, gravel and gravel), which ranks 1st in quantity and 2nd in value in the world. The low unit prices result from the relationship between distribution distance limits (local use) and the wide distribution of small enterprises.

It is used in the construction industry to make concrete, surfacing for dirt roads, paving, cyclopean concrete, garden ornamentation, etc. Naturally occurring throughout the earth's crust, they are found mainly in rivers, and sometimes in layers of sedimentary rocks with low cohesion, weathering of crystalline rocks, or resulting from the processing of sand (ALECRIM, 1982; OLIVEIRA, 2002).

The use of gravel to make bases and sub-bases has always been viable because it is available in many regions, at low or even no cost to extract or exploit, and also because it gives a good end result. With the optimisation and increased weight of the means of transport, there is a need for pavements with greater resistance. With this high demand, it has become increasingly difficult to find gravel that meets the basic specifications (OLIVEIRA, 2012).

A key factor in defining the properties and behaviour under the action of traffic is the external shape of the minerals, which affects the effectiveness of their role in paving, with cubic or spherical grains being the best to use compared to other existing grains. This is because other grains break easily, thus altering the control carried out throughout the paving project, according to Senço (2008).

Because it was a material that easily met traditional specifications, its use in paving works began around 65 years ago, given that when compacted it had a high bearing capacity (California Bearing Ratio - CBR) (greater than 80 per cent) and could even be used on roads with a high volume of traffic.

According to Cortez, the use of materials for civil construction has been going on for some time, mostly due to the high rate of urbanisation, which has led to the country's biggest urban demographic growth in the last two decades.

As a result, there has been a marked increase in recent years in construction activities, with the consequent participation of extracted materials, both for direct use, such as sand and pebbles, as well as other materials, such as clay, used for industrialisation.

The growing need for these materials, coupled with various factors such as the search for quick and easy profits, the free and simple obtaining of raw materials, the lack of environmental awareness among those involved, and finally the total absence and even complicity of the official bodies responsible for overseeing these activities, has led to accelerated environmental degradation in countless areas.

According to Article 1 of CONAMA Resolution 001/86, an environmental impact is considered to be any alteration to the physical, chemical and biological properties of the environment, caused by any form of matter or energy resulting from human activities that directly or indirectly affect it:

I - The health, safety and well-being of the population;

II - Social and economic activities;

III - The biota;

IV - The aesthetic and sanitary conditions of the environment;

V - The quality of environmental resources.

Therefore, these human activities also refer to the extraction of any natural material. These precautions must be taken in the process of removing gravel responsibly, not just for economic reasons, so that neither party loses out and these impacts are minimised.

According to GROSSI & VALENTE (2013), the aim of a mineral resource classification system is to apply rules that make it possible to separate different classes of resources in such a way that part of them can be transformed into reserves, based on technical and economic feasibility studies. With this information, it is possible to guide decision-making to maximise the efficiency with which this resource is used.

Reserves are reliable when, once determined, they make it possible to calculate admissible errors for each established class. Calculating this error requires good quality data in adequate quantity for the specific case under study.

In the municipality of Araxá-MG, the volume of gravel reserves suitable for use in paving has not been registered and calculated, so it is of fundamental importance to carry out such a survey. Therefore, this study aims to locate and size gravel reserves suitable for use in paving and road rehabilitation works.

Because they are easily accessible areas, the Serrinha, Marmelo and Itaipu regions (areas with a large population), located in the municipality of Araxá-MG, make it possible to exploit the gravel deposits found there.

CHAPTER 2

OBJECTIVE

- Registering and calculating the volume of gravel reserves suitable for use in paving in the Serrinha, Marmelo and Itaipu regions, extensions located in Araxá, in Alto Paranaíba, Minas Gerais. .

- Georeference and register the identification points of the gravel used in the research.

CHAPTER 3

MATERIALS AND METHODS USED

The materials used in the research were obtained from the regions of Itaipu (FIGURE 1) and Marmelo (FIGURE 2), located in the municipality of Araxá-MG, as shown in Figure 3, with each sample georeferenced and the area of the gravel deposit calculated.

Figure 1- Itaipu region in the municipality of Araxá/ MG. Source: Taken by the authors.

Figura 2-   Marmelo region in the municipality of Araxá/MG.
Source: Taken by the authors.

Figura 3-   Araxá Road System.
Source: IPDSA, 2017.

In the Serrinha region -municipality of Araxá- sampling was not

carried out due to

the difficulties encountered in carrying it out, but the experiment

continued with the other regions.

The data collection points were selected in such a way as not to impact the environment, from no deforestation to drastic alteration of the physical environment situated as shown in Figures 4 and 5, extracting a total of approximately eight kilos of samples at each point.

Figure 4- Physical means of sample extraction. Source: Personal collection.

Figure 5- Sample extraction starting point. Source: Personal collection.

Once the samples had been collected, they were stored in suitable containers and labelled according to their location and quantity, as shown in the figures below.

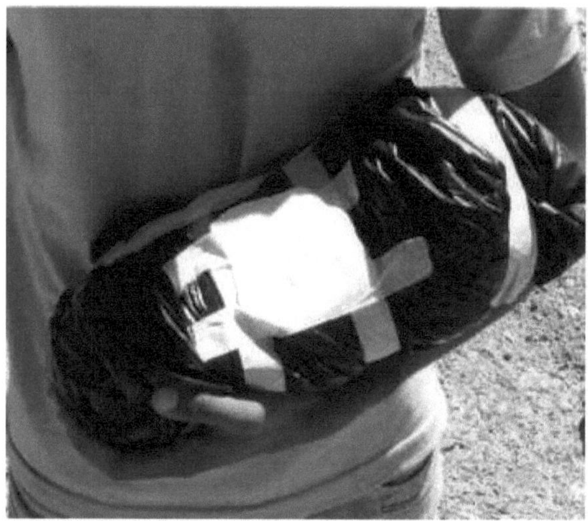

Figure 6- Sample from the Marmelo region.
Source: Personal collection

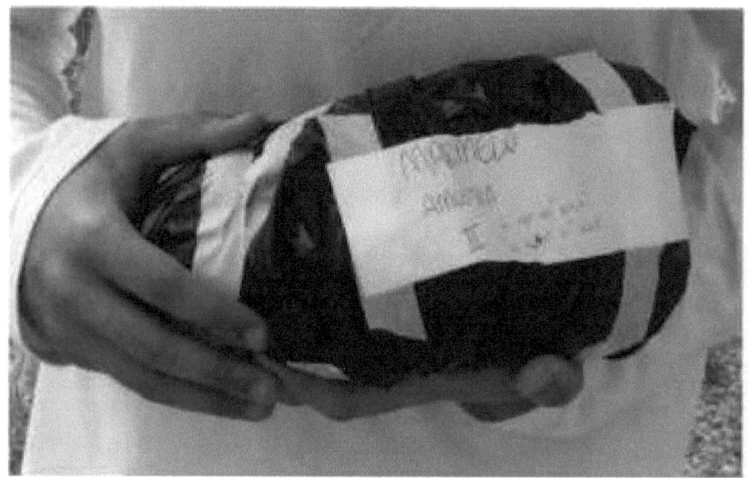

Figura 7-   Sample from the Marmelo region
Source: Personal collection.

Figura 8-   Sample from the Itaipu region.
Source: Personal collection.

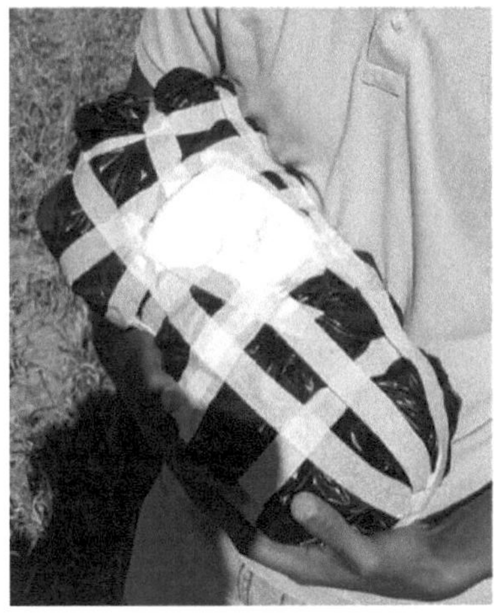

Figure 9- Sample from the Itaipu region. Source: Personal collection.

The sample preparation stage for laboratory tests began with the quartering and reduction of the material, in accordance with specification NBR NM 27 (ABNT, 2001).

Initially, the classification test for the gravel was carried out in accordance with Annex A of NBR 15116 (ABNT, 2004). Afterwards, the particle size test was carried out for each aggregate fraction by sieving, in accordance with DNER-ME 083/98 (1998). In this test, three repetitions were made, so that an average particle size curve was obtained, emphasising that all processes were carried out for all regions. In addition, the maximum characteristic size, the uniformity coefficient and the percentage passing the 0.42 mm sieve (No. 40) were determined for the three regions.

Using the results obtained from the particle size test, the particle size correction was carried out using the trial method, with the aim of obtaining a mixture that was as close as possible to band C of DNER - ES 303/97 (1997).

CHAPTER 4

RESULTS AND DISCUSSIONS

Geo-referencing is essential for locating gravel deposits, as it makes it possible to create a control point that offers a perfectly identifiable physical feature.

The deposit with the largest volume of gravel and feasibility for exploitation in the Itaipu region (coordinates: Latitude: -19.615262° and Longitude: -47.165806°), is located within the polygon formed by coordinates 23K 026844;UTM 7834160, 23K 026839;UTM 7834228, 23K 026833;UTM 7834171, 23K 26838;UTM 7834129, with an estimated area of 1480m^2 , gross volume of 669.70m^3 , and gravel volume of 321.87m3.

The deposit found in the Marmelo region (coordinates 19°35'39"S and 047°04'52.4"W) was georeferenced as shown in Figure 2 and has an area of approximately 710m^2 , containing an approximate volume of 404.7 m^3 .

Georeferencing of collection points

Sample I	Sample II
S 19°38"03,8'	S 19°38"04,4'
W 047°10"59,5'	W 047°11"00,3'
Depth: 0.54m	Depth: 0.37 metres
Sample III	**Sample IV**
S 19°38"03,8'	S 19°38"02,6'

W 047°10"58,9'	W 047°10"58,3'
Depth: 0.59m	Depth: 0.51 metres

Figure 10- Georeferencing of the collection points in the Marmelo region.
Source: Prepared by the authors.

The samples were quartered, fractionated and reduced, using only a quarter of each portion.

The gravels from both regions have similar characteristics and for the classification of the gravel from the two extensions, it was observed that 93% of its mass was retained up to the 4.8mm sieve, so it could be classified as coarse aggregate.

The particle size curve gives an idea of the shape of the curve, allowing us to detect discontinuities in the whole. Soil can be well graded, when its size gradually decreases, uniformly graded soil, with uniformity in size, or openly graded soil, when its size gradually decreases and has a considerably high decrease interval.

The external shape of aggregates is an important factor in defining their properties and behaviour under the action of traffic, as it is clear that a cubic or spherical grain behaves better than an angular or flat grain. The presence of easily broken grains can lead to a total change in the grain size of an aggregate, rendering all the control work carried out during the design and construction of the pavement useless, as mentioned by Senço (2008). The graphs below show the grain size curve of the gravel samples analysed.

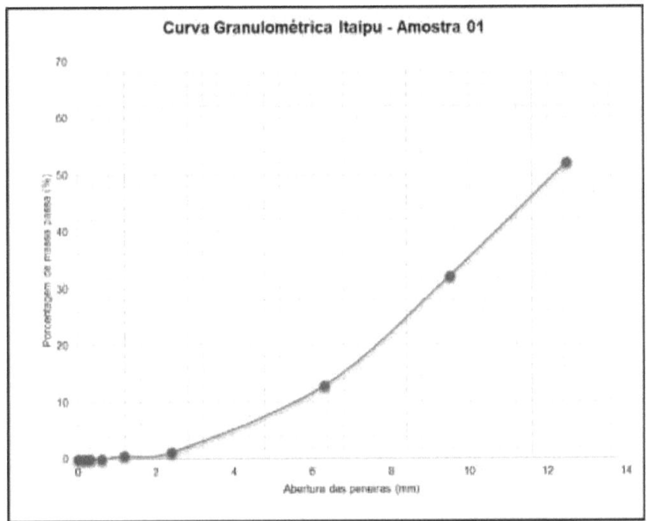

Figure 11- Particle size curve of Itaipu | Sample 1. Source: Prepared by the authors.

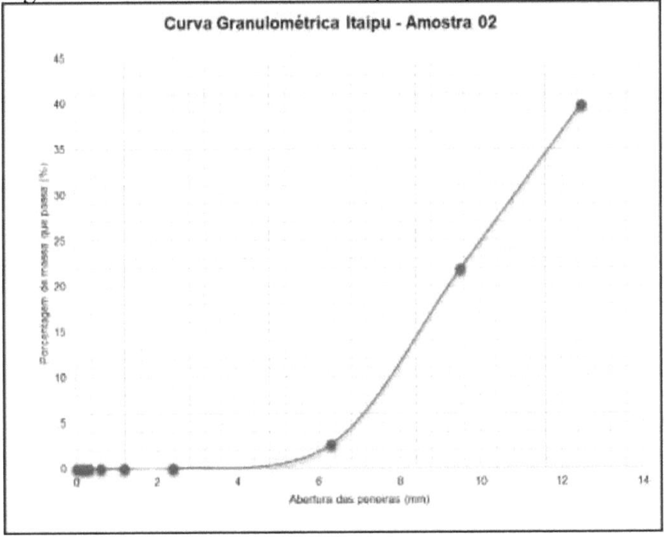

Figura 12- Particle size curve of Itaipu | Sample 2.
Source: Prepared by the authors.

21

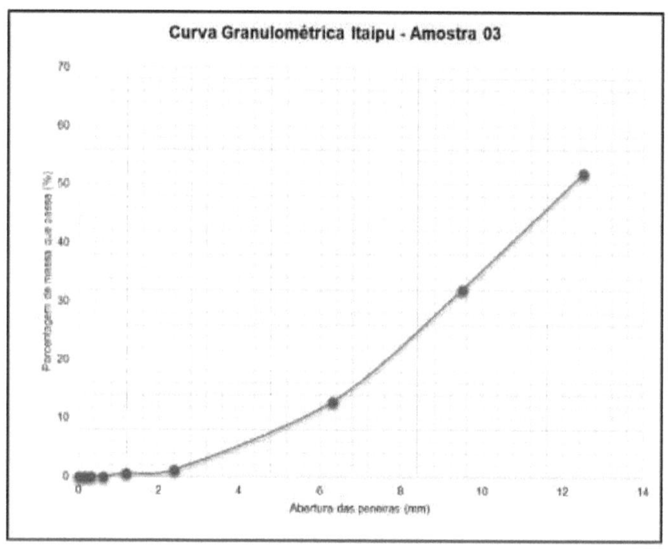

Figura 13-   Particle size curve of Itaipu | Sample 3.
Source: Prepared by the authors.

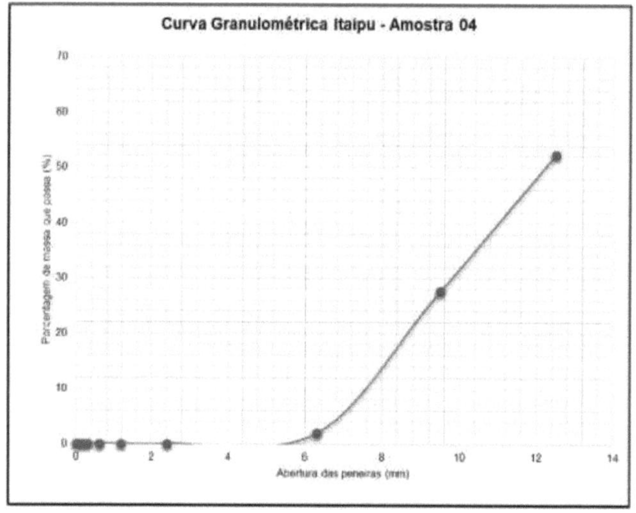

Figura 14-   Particle size curve of Itaipu | Sample 4.
Source: Prepared by the authors.

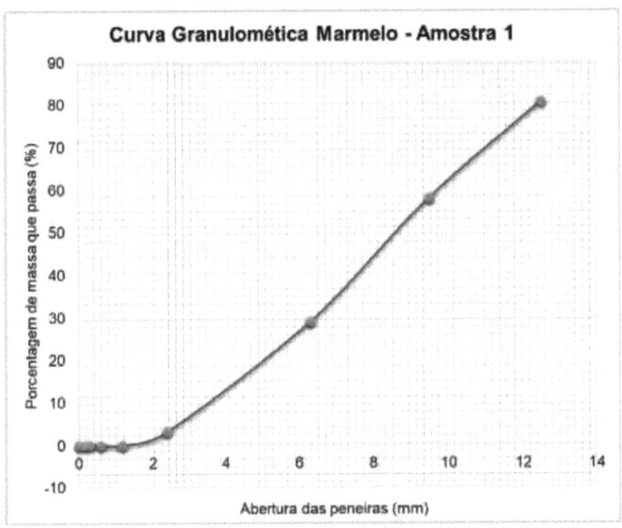

Figura 15- Quince granulometric curve | Sample 1.
Source: Prepared by the authors.

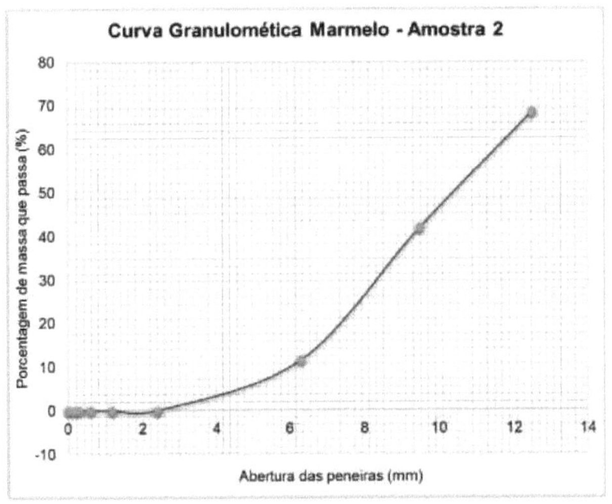

Figura 16- Quince granulometric curve | Sample 2.
Source: Prepared by the authors.

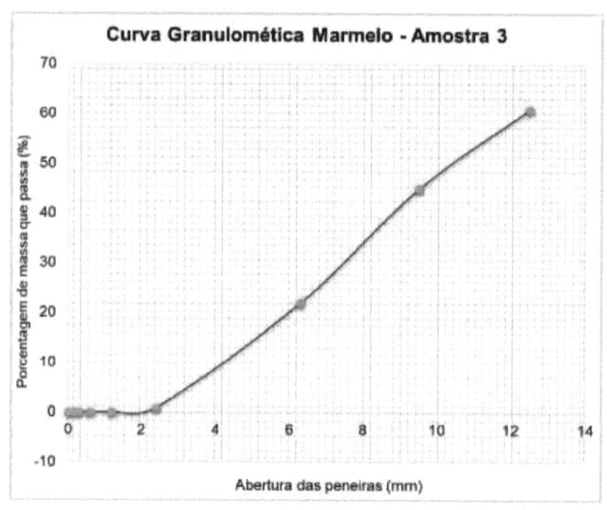

Figura 17- Quince granulometric curve | Sample 3. Source: Prepared by the authors.

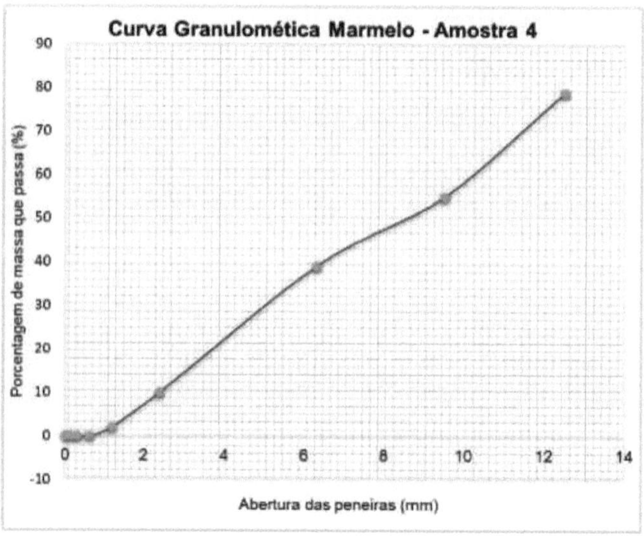

Figure 18- Marmelo particle size curve | Sample 4. Source: Prepared by the authors.

After drawing up the graphs above, the results were compared with the classification proposed by Varela (2012), shown in Figure 11 below:

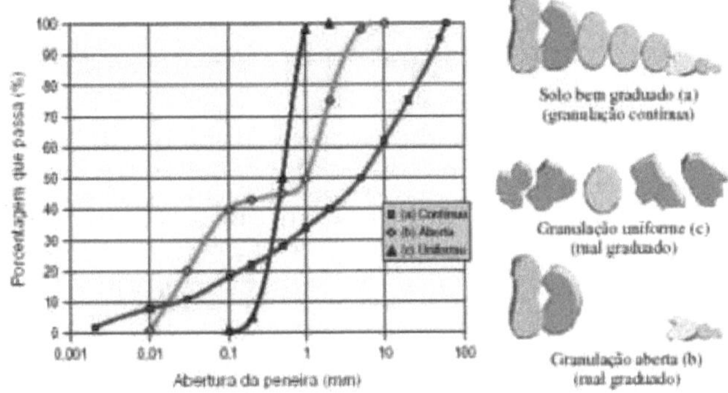

Figure 19- Classification of aggregate granulation.

Source: Varela, 2012

It can be seen that all the gravel samples analysed have a well graded soil classification.

Figure 12, which shows the maximum diameter of the coarse aggregate, was drawn up on the basis of analysing the mesh of the sieve in which the percentage equal to or immediately less than 5% was retained (VARELA, 2012).

MAXIMUM CHARACTERISTIC DIAMETER

Region	Gravel samples	DMC (mm)
Itaipu	Sample 01	19
	Sample 02	19
	Sample 03	19
	Sample 04	19
Quince	Sample 01	19
	Sample 02	19
	Sample 03	19
	Sample 04	19

Figure 20- Maximum characteristic diameter of the gravel samples analysed.
Source: Prepared by the authors.

Comparing the properties of the gravel present in each sample, a similarity was observed between each fraction analysed.

With the increase in the use of this mineral and, consequently, the demand for greater technical knowledge, the concept of quality is increasingly being associated with economic aspects.

CHAPTER 5

CONCLUSION

The gravel located in the Marmelo and Itaipu regions is suitable for use in paving, such as the construction of new roads, urban roads or the maintenance of rural roads.

Protocols involving a greater variety of mechanical tests and a larger sample size are suggested for future studies, so that the material's properties can be better evaluated.

CHAPTER 6

BIBLIOGRAPHICAL REFERENCES

ALECRIM, José Duarte. **Mineral resources of the state of Minas Gerais**. Belo Horizonte: METAMIG. 1982.

BRAZILIAN ASSOCIATION OF TECHNICAL STANDARDS. NBR 15116: Recycled aggregates from solid construction waste - Use in paving and concrete preparation without structural function - Requirements. Rio de Janeiro: Abnt, 2004. 12 p.

BRAZILIAN ASSOCIATION OF TECHNICAL STANDARDS. NBRNM 27: Aggregates: Field sample reduction for laboratory tests. Rio de Janeiro: Abnt, 2001. 07 p.

BRAZILIAN ASSOCIATION OF TECHNICAL STANDARDS. **NBR 9895**: Soil - California Support Index. Abnt, 1987. 14 p.

BAUER, Luis Alfredo Falcão. **Construction Materials: New Materials for Civil Construction.** 5. ed. São Paulo: Ltc, 2013. 471 p.

NATIONAL ENVIRONMENTAL COUNCIL. **CONAMA RESOLUTION 001/86**. Brasilia, 1986.

NATIONAL DEPARTMENT OF MOTORWAYS. DNER-ME 081/98: Aggregates - determination of absorption and density of coarse aggregate. Rio de Janeiro: Dner, 1998. 6 p.

NATIONAL DEPARTMENT OF MOTORWAYS. DNER-ME 083/98: Aggregates - particle size analysis. Rio de Janeiro: Dner, 1998. 10 p.

NATIONAL DEPARTMENT OF MOTORWAYS. DNER-ES

303/97: Paving - Stabilised Base

Granulom et rica mente. Rio de Janeiro: Dner, 1997. 7 p.

NATIONAL DEPARTMENT OF TRANSPORT INFRASTRUCTURE. DNIT 098/2007 - ES: Paving - Granulometrically Stabilised Base using Lateritic Soil - Service Specification. Rio de Janeiro, 2007.

GROSSI, J.; VALENTE, Jorge. Practical Guide to Calculating Mineral Resources and Reserves. Brasília, 2003.

IPDSA (Araxá Institute for Planning and Sustainable Development). **Municipal collection of topographic maps.** 2017.

MARANGON, Márcio. Topics in Geotechnics and Earthworks. 2009. 14 f. Dissertation (Master's) - Civil Engineering Course, Federal University of Juiz de Fora, Juiz de Fora, 2011.

OLIVEIRA, Murilo Rassi de. Analysing aggregates for use in paving. 2012. 1 v. TCC (Graduation) - Civil Engineering Course, Pontifical Catholic University of Goiás, Goiânia, 2012.

SENÇO, Wlastermiler de. Manual of Paving Techniques. 2. ed. São Paulo: PINI, 2008.

VARELA, Marcio. Granulometry. São Paulo, 2012. 34 slides, colour.

"lmmb-s-sq that people can take everything from you except your knowledge. "

\- Albert Einstein

I want morebooks!

Buy your books fast and straightforward online - at one of world's fastest growing online book stores! Environmentally sound due to Print-on-Demand technologies.

Buy your books online at
www.morebooks.shop

Kaufen Sie Ihre Bücher schnell und unkompliziert online – auf einer der am schnellsten wachsenden Buchhandelsplattformen weltweit! Dank Print-On-Demand umwelt- und ressourcenschonend produziert.

Bücher schneller online kaufen
www.morebooks.shop

info@omniscriptum.com
www.omniscriptum.com

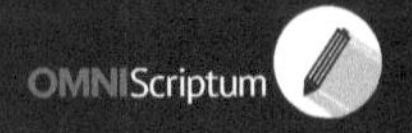

Printed by Books on Demand GmbH, Norderstedt / Germany